÷ mini maths +

Patterns

David Kirkby

answers

Page 5 The cushion, T-shirt and cup have stripes going across.
The wash-bag and towel have stripes going up and down

Page 7 Spots

Page 9 Red, yellow, red, yellow

Page 11 32 black squares, 32 white squares

Page 15 Yes

Page 17 Yes

Page 19 Spots, checks, stripes, picture patterns, diamonds

Page 21 6, 8

Page 23 Anywhere that cuts through the centre

---

index

Check patterns 10, 11
Colour patterns 8, 9
Number patterns 20, 21
Picture patterns 16, 17
Solid patterns 12, 13
Spots 6, 7
Stacking patterns 12, 13
Stripes 4, 5
Symmetrical patterns 22, 23
Tessellating patterns 14, 15